YOLO COUNTY LIBRARY

WITHDRAWN

Friends of the
Davis Public Library

Providing Books
and Programs

Our Universe

Saturn

by Margaret J. Goldstein

Lerner Publications Company • Minneapolis

Text copyright © 2003 by Margaret J. Goldstein

All rights reserved. International copyright secured. No part of this book may be reproduced, stored in a retrieval system, or transmitted in any form or by any means—electronic, mechanical, photocopying, recording, or otherwise—without the prior written permission of Lerner Publications Company, except for the inclusion of brief quotations in an acknowledged review.

Lerner Publications Company
A division of Lerner Publishing Group
241 First Avenue North
Minneapolis, MN 55401 USA

Website address: www.lernerbooks.com

Words in **bold type** are explained in a glossary on page 30.

Library of Congress Cataloging-in-Publication Data

Goldstein, Margaret J.
    Saturn / by Margaret J. Goldstein.
      p. cm. — (Our universe)
    Includes index.
    Summary: An introduction to Saturn, describing its place
in the solar system, its physical characteristics, its
movement in space, and other facts about this planet.
    ISBN: 0-8225-4653-1 (lib. bdg. : alk. paper)
    1. Saturn (Planet)—Juvenile literature. [1. Saturn (Planet)]
I. Title. II. Series.
QB671 .G66 2003
523.46—dc21   2002000947

Manufactured in the United States of America
1 2 3 4 5 6 — JR — 08 07 06 05 04 03

The photographs in this book are reproduced with permission from: NASA, pp. 3, 5, 12, 13, 14, 15, 17, 19, 20, 21, 22, 23, 25, 26; © Finley Holiday Films/Photo Network, p. 9; © John Sanford/Photo Network, p. 27.

Cover: NASA.

This yellow planet is famous for its wide, flat rings. What planet is it?

The planet is Saturn. Saturn is a giant planet. It is much larger than Earth. More than 750 planets as big as our planet could fit inside Saturn.

Earth

Saturn

Saturn is one of the largest planets in the **solar system.** The solar system has nine planets in all. Only Jupiter is larger than Saturn.

All of the planets in the solar system **orbit** the Sun. To orbit the Sun means to travel around it. Saturn orbits the Sun in an oval path. It is the sixth planet from the Sun.

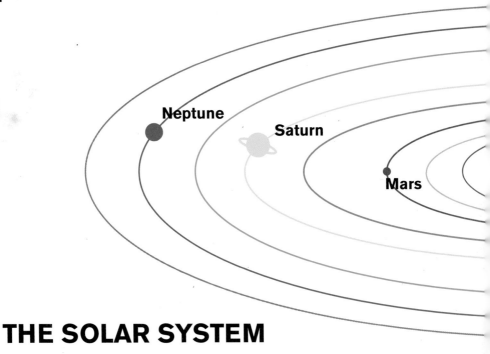

THE SOLAR SYSTEM

Saturn takes about 29 years to orbit the Sun once. Earth's trip is much shorter. It orbits the Sun in 1 year.

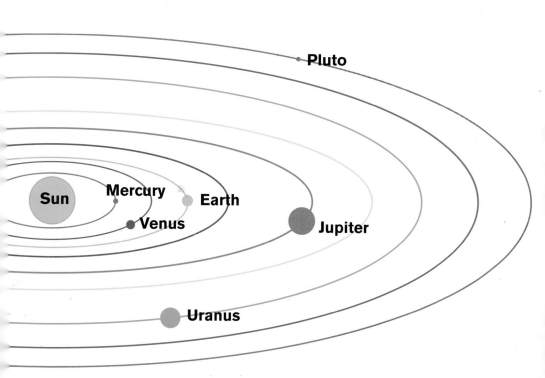

Planets also spin around as they travel through space. This kind of spinning is called **rotating.**

Saturn rotates quickly. One full spin takes about 11 hours. Only Jupiter rotates faster than Saturn. Earth takes 24 hours to rotate once.

Saturn is surrounded on the outside by a thick layer of gases. This layer is called an **atmosphere.**

Saturn is not a solid planet like Earth is. Under the atmosphere are two thicker liquid layers. The very center of Saturn may be a hard ball of rock.

SATURN'S LAYERS

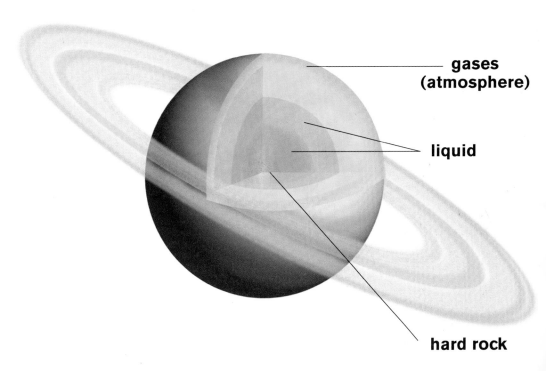

It is very cold at the top of Saturn's atmosphere. But Saturn is warmer below its atmosphere. The planet grows even hotter deeper inside. Saturn's center is probably burning hot.

Saturn's atmosphere is full of colorful clouds. Fast winds whip these clouds around the planet. The winds also cause powerful storms.

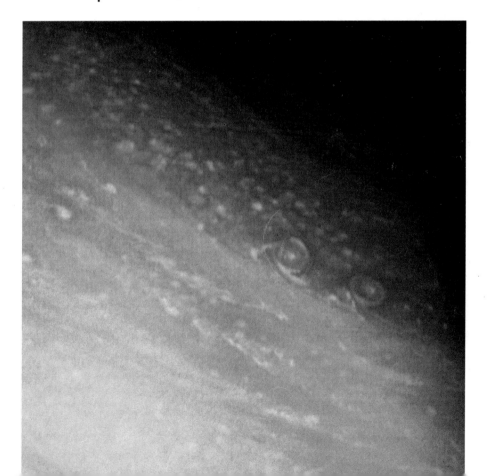

Big, wide rings orbit Saturn. The rings reach far into space. Jupiter, Uranus, and Neptune also have rings. But Saturn has the largest set of rings in the solar system.

Saturn has seven rings in all. The rings are made up of smaller rings called ringlets. What are the ringlets made of?

The ringlets are made mostly of billions of pieces of ice. Some of the pieces are as small as dust. Other pieces are as big as boulders.

An artist made this picture of the bits of ice in Saturn's rings.

Something else orbits Saturn along with the rings. At least 30 moons orbit the planet.

Some of Saturn's moons orbit in between the rings. Other moons orbit farther away from the planet.

Saturn with its moons Tethys and Dione

Some of Saturn's small moons

Saturn has six large moons. These moons are round like Earth's moon. Saturn's other moons are smaller. Many of the smaller moons look like large rocks.

Titan is Saturn's largest moon. It is bigger than Earth's moon. It is even bigger than the planet Mercury. Titan has a thick orange atmosphere.

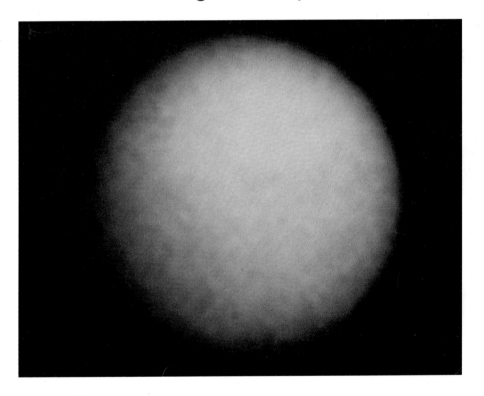

In 1973, Americans sent a spacecraft to Saturn for the first time. The spacecraft was named *Pioneer 11*.

Two other spacecraft were sent to Saturn in the 1980s. These spacecraft were called *Voyager 1* and *Voyager 2*.

An artist made this picture of *Voyager 2* orbiting Saturn.

23

The *Pioneer* and *Voyager* spacecraft studied Saturn's atmosphere. They took pictures of the planet. They took pictures of its rings and moons. The spacecraft found rings and moons that people had not known about before.

An artist made this picture of *Cassini*.

Another spacecraft is on its way to Saturn. It left Earth in 1997. The name of the spacecraft is *Cassini.* It is due to reach Saturn in 2004.

Look into the sky at night. Can you see Saturn? What do you think you would find if you could go there?

Facts about Saturn

- Saturn is 887,000,000 miles (1,430,000,000 km) from the Sun.
- Saturn's diameter (distance across) is 74,900 miles (120,000 km).
- Saturn orbits the Sun in 29 years.
- Saturn rotates in 11 hours.
- The average temperature in Saturn's atmosphere is −218°F (−140°C).
- The atmosphere of Saturn is made of hydrogen and helium.
- Saturn has at least 30 moons.
- Saturn has 7 rings.
- Saturn was named after the Roman god of agriculture.

- Saturn has been visited by *Pioneer 11* in 1979, *Voyager 1* in 1980, and *Voyager 2* in 1981. *Cassini* will reach Saturn in 2004.

- Winds on Saturn can blow more than 1,000 miles (1,600 km) an hour.

- Saturn is nine times wider than Earth.

- Saturn is called a "gas giant" because it is very big and is made mostly of gas.

- If you could put Saturn in water, it would float.

- Saturn's moon Mimas has a big crater that looks like a giant eyeball.

Glossary

atmosphere: the layer of gases that surrounds a planet or moon

orbit: to travel around a larger body in space

rotating: spinning around in space

solar system: the Sun and the planets, moons, and other objects that travel around it

Learn More about Saturn

Books

Brimner, Larry Dane. *Saturn.* New York: Children's Press, 1999.

Simon, Seymour. *Saturn.* New York: Morrow, 1998.

Websites

Solar System Exploration: Saturn
<http://solarsystem.nasa.gov/features/planets/saturn/saturn.html>
Detailed information from the National Aeronautics and Space Administration (NASA) about Saturn, with good links to other helpful websites.

The Space Place
<http://spaceplace.jpl.nasa.gov>
An astronomy website for kids developed by NASA's Jet Propulsion Laboratory.

StarChild
<http://starchild.gsfc.nasa.gov/docs/StarChild/StarChild.html>
An online learning center for young astronomers, sponsored by NASA.

Index

atmosphere, 10, 11, 12–13, 24, 28

Cassini spacecraft, 26, 29

clouds, 13

distance from the Sun, 28

layers, 10, 11

moons, 18, 19, 20–21, 24, 29

orbit, 6–7, 28

Pioneer 11 spacecraft, 22, 24, 29

rings, 3, 14–17, 18, 24, 28

rotation, 8, 28

size, 4–5, 28, 29

Sun, 6

solar system, 5–7

spacecraft, 22–24, 26, 29

temperature, 12, 28

Titan, 21

Voyager spacecraft, 23–24, 29

winds, 13, 29

MAR 1 4 2005 DA